T0361531

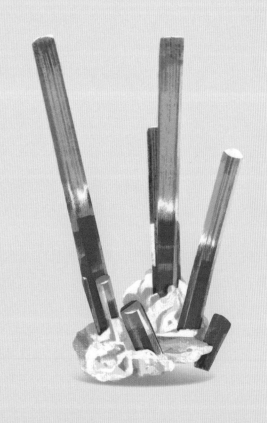

Phaidon Press Inc.
111 Broadway
New York, NY 10006

First published in 2024
© 2024 Phaidon Press Limited
Text © 2024 Maya Wei-Haas
Illustrations © 2024 Sonia Pulido

Artwork created digitally

ISBN 978 1 83866 790 0 (US edition)
006-1223

A CIP catalog record for this book is
available from the Library of Congress.
All rights reserved. No part of this publication
may be reproduced, stored in a retrieval
system or transmitted, in any form or
by any means, electronic, mechanical,
photocopying, recording or otherwise,
without the written permission of
Phaidon Press Limited.

Edited by Maya Gartner
Designed by Kevin Knight
Production by Rebecca Price

Printed in China

What a Rock Can Reveal

Where They Come From
And What They Tell Us
About Our Planet

BY **MAYA WEI-HAAS**

ILLUSTRATED BY **SONIA PULIDO**

There's no such thing as a boring rock!

Some are flashy with pops of pink or glitters of green. Others have swirls and stripes or spots of all sizes. They can be rough or smooth, thin or wide. They can even be sneaky and hide treasure inside.

Even rocks you think look plain, like a little gray pebble with no sparkle or shine, has gone on a big adventure. Some come from deep underground, while others have raced through space.

So don't let a simple stone trick you. Every rock you'll ever find has an exciting story to tell. Let's discover how to read them!

Next time you stumble on a stone, take a close look

What can you see? Colorful polka dots? Layers that twist and turn? Tiny holes all over? The way a rock looks can tell us where it comes from and what adventures it's had along the way.

Gneiss

Basalt with olivine crystals

Chert

Quartzite

Pink granite

Conglomerate

Gabbro

Pumice

Marble

Banded rhyolite

Lapis lazuli

Granite

Eclogite

Limestone with ammonite fossils

Gneiss

Lherzolite

Shale

Sandstone

Unakite

Soapstone

Obsidian

What can a rock's texture reveal?

Rub your fingers across the next rock you see. What do you feel? Is it rough with points and ridges? It may have broken apart not too long ago. Is it smooth and round? It may have gone on an epic journey.

Wind and water can slowly wear big boulders down, sometimes making them weak enough to crack. Rocks can also crumble as they roll downhill, even crushing other chunks as they fall.

Rocks that carry on with their adventure get smoother the farther they go. They tumble down streams, bumping and rubbing together turning rough rocks into polished pebbles. You can find them on beaches and near rivers, or anywhere water flows!

What can a rock's colors reveal?

Rocks are made from a rainbow of beautiful building blocks called minerals.

Minerals grow in nature, often underground, and come in crystals of all different colors and shapes. Some rocks are made from just one mineral. Others have a mix of many minerals that you can see as tiny speckles or big spots.

Garnet

Omphacite

Kyanite

Quartz

ECLOGITE

This special rock is very hard to find. Each color—bright green and red and sometimes specks of blue and white—is a different mineral. Which one do you like best?

GRANITE

Granite is a gray or pink rock that can be found all over the world. Just like eclogite, it's covered in spots. Each patch of color is a different mineral.

Hornblende

Muscovite

Orthoclase feldspar

Smoky quartz

Plagioclase feldspar

Let's meet some minerals!

Open a treasure chest and you might find a rainbow of sparkling minerals hiding inside— emeralds, rubies, sapphires, and diamonds galore! For thousands of years these gems were used in crowns and necklaces for kings and queens.

VIVIANITE

This trickster can be clear as glass when first pulled from the ground. But light makes it slowly turn dark blue, green, or even black.

DIAMOND

They dazzle and shine in expensive jewelry, but diamonds are also so hard they can be used to drill into solid rock.

MAGNETITE

One type of this mineral has a curious talent: it can attract many metals, making it a natural magnet! You can find bits of magnetite by swishing a magnet in sand.

CROCOITE

Is it a cactus? No, but this hard-to-find, bright-red mineral is super spiky. You've been warned!

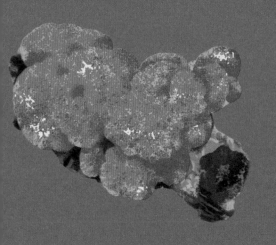

RHODOCHROSITE

This collection of crystals is an exciting and unusual way for this mineral to form. While it might look like bubble gum, it's way too hard to chew!

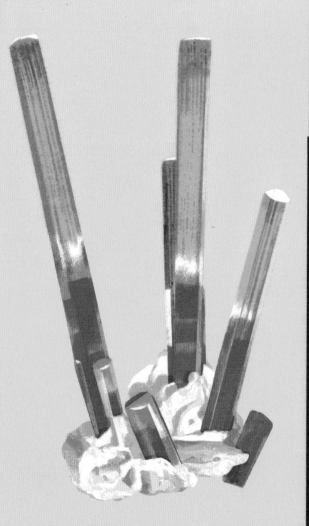

TOURMALINE

This mineral grows as tall sticks in spectacular colors that make it look like candy. One type even has the bright pinks and greens of a delicious watermelon slice.

But minerals aren't just pretty; they can be useful too. Quartz is used in clocks, mica makes many paints sparkle, and calcite is used in chalk! We've found thousands of minerals, but there are many more left to discover.

CALCITE
This one has a bubbly personality. Add a few drops of lemon juice or vinegar, and it fizzes!

AZURITE
Ancient Egyptians and famous painters like Leonardo da Vinci loved to use this bright blue mineral to make beautiful paint or dyes.

SALT
Did you know that the salt you put on your food is a mineral? It grows in itty-bitty cube-shaped crystals.

FLUORITE
This mineral glows neon blue under a black light. Some kinds even keep shining for a while after you turn off the light.

Some minerals like hot, cramped places deep underground, while others sit on Earth's surface. Minerals can mix in many ways that give us clues about where they came from. What minerals have you seen? Can you guess their stories?

Where can you find rocks?

Rocks are everywhere! You can find them in parks near the flowers and trees, or in playgrounds under slides and swings. You might even find them outside your front door, beneath leaves or along garden paths. The more you look around, the more rocks you'll see.

Even if you don't spot many stones, there's always rock deep under your feet—whether you're hiking up a mountain or strolling down a street. That's because Earth is mostly made of rock.

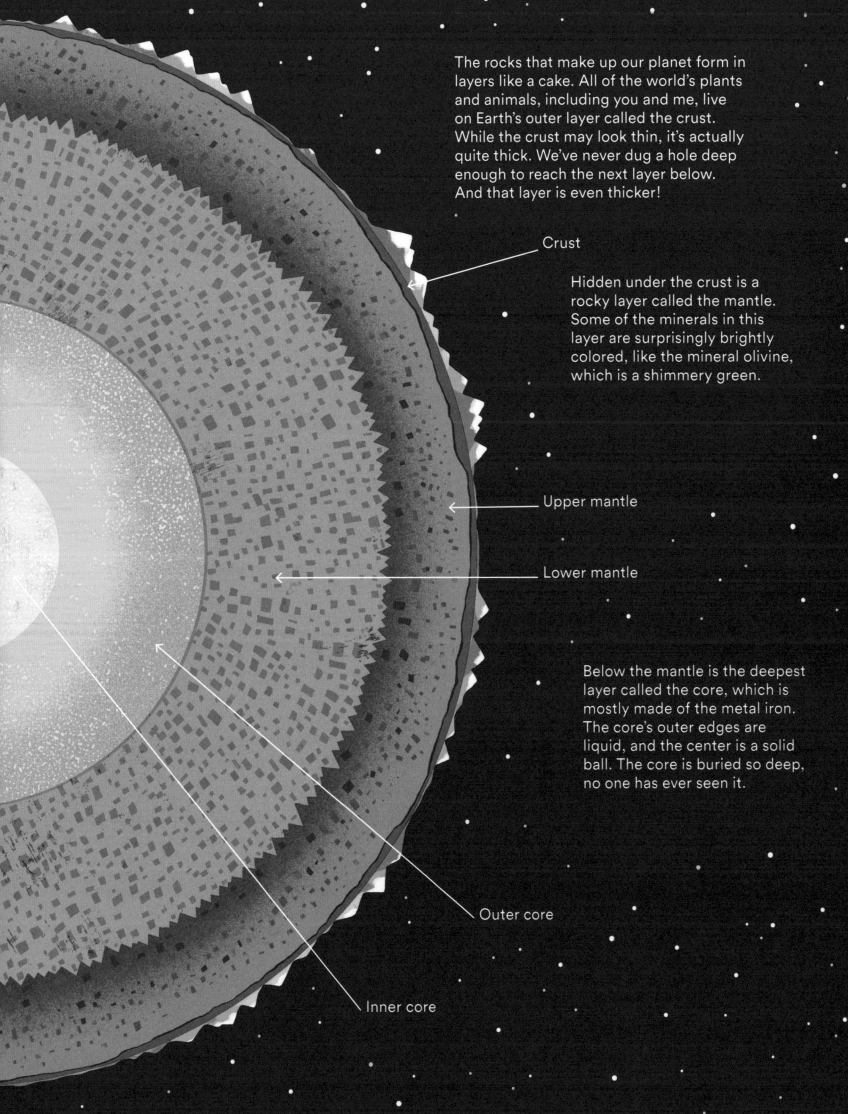

The rocks that make up our planet form in layers like a cake. All of the world's plants and animals, including you and me, live on Earth's outer layer called the crust. While the crust may look thin, it's actually quite thick. We've never dug a hole deep enough to reach the next layer below. And that layer is even thicker!

Crust

Hidden under the crust is a rocky layer called the mantle. Some of the minerals in this layer are surprisingly brightly colored, like the mineral olivine, which is a shimmery green.

Upper mantle

Lower mantle

Below the mantle is the deepest layer called the core, which is mostly made of the metal iron. The core's outer edges are liquid, and the center is a solid ball. The core is buried so deep, no one has ever seen it.

Outer core

Inner core

Some rocks tell stories of big blasts and booms

Have you ever found a completely black rock? It might be basalt. This rock looks plain but comes from one of Earth's most amazing events: a volcanic eruption that oozes super-hot, melted—molten—rock, also known as lava.

Some lava can flow like rivers, glowing bright yellow, orange, and red.

As the lava cools, its glow fades and darkens to form solid rock, like the black rock basalt.

Volcanic blasts can fling lava sky high, lighting up the night like nature's fireworks. Some eruptions are so powerful they can knock over trees and "boom" so loud you can hear them from miles away.

IGNEOUS ROCKS

Rocks that cool from magma or lava are called igneous rocks. You can find one type of igneous rock called basalt all over the world.

HOW DO VOLCANOES ERUPT?

It all starts deep underground when small bits of Earth's mantle and crust melt to form magma, which is a sizzling-hot stony soup. This "soup" oozes upward and collects in super-hot blobs that press on the ground above. Suddenly, the surface splits with a booming blast. That's an eruption!

How do volcanic rocks show their adventures?

A volcanic rock's shape, texture, size, and more can give exciting clues about what happens after it blasts up from the ground.

LAVA BOMBS
Watch out! Flung high up out of a volcano, these molten blobs smoosh into a football shape as they soar through the sky.

PELE'S TEARS
Don't worry, the volcano isn't crying. When tiny globs of lava shoot into the air, they quickly cool to form these droplets of volcanic glass. They're named after Pele, the Hawaiian goddess of volcanoes.

PAHOEHOE
Pronounced "pah-hoy-hoy," this rock forms from lava that slowly flows across the ground, twisting and crossing like ropes.

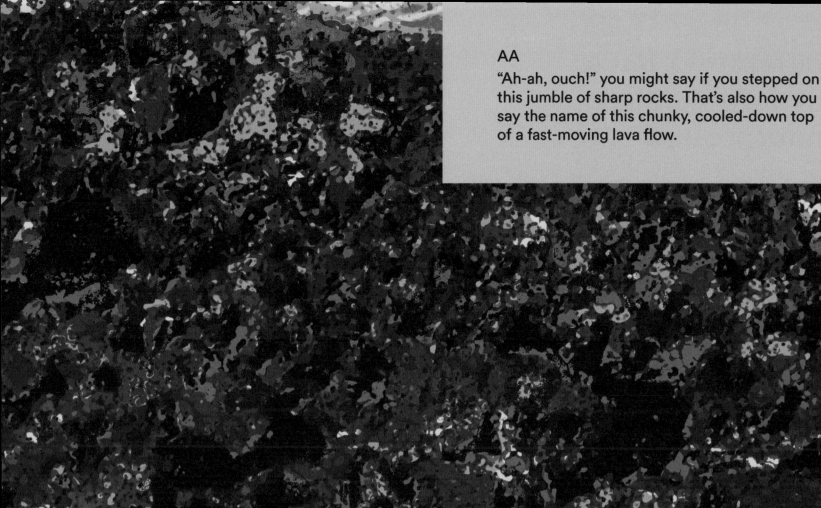

AA

"Ah-ah, ouch!" you might say if you stepped on this jumble of sharp rocks. That's also how you say the name of this chunky, cooled-down top of a fast-moving lava flow.

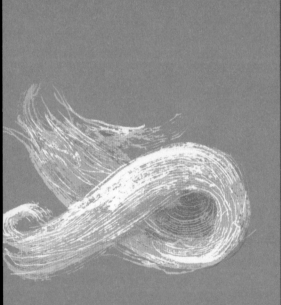

PELE'S HAIR

If you're lucky, you just might find these thin, delicate strands that form when lava is thrown into the air. The gooey rock stretches and quickly cools, making wispy threads of volcanic glass. Sometimes they're even attached to one of Pele's tears!

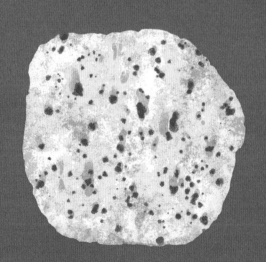

PUMICE

Just like the fizz from a freshly shaken soda, erupting volcanoes can spit up lava with lots of gas. This makes rocks that are full of holes—so many that pumice can float on water.

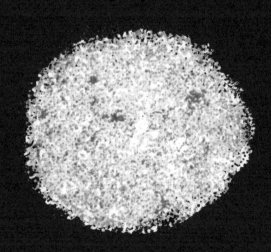

VOLCANIC ASH

These are tiny, sharp pieces of broken-up lava rock from the blast of an erupting volcano. During big explosions, ash rains from the sky, even if the volcano is miles away.

Some rocks have adventures deep underground

Many rocks form deep beneath your feet. Some of these stones start as blobs of liquidy magma that get stuck underground. The molten rock slowly cools, which makes the perfect environment for minerals to grow. These crystals stacks up like toy blocks to create solid rocks, like granite!

Gold

Gypsum crystal cave

Other sparkly minerals need a trickle of water and heat to turn into something neat. The water seeps into underground cracks, dissolving some minerals, kind of like mixing sugar with piping-hot water. As the liquid cools, new mineral crystals grow. They fill nooks and crannies with a rainbow of minerals and gems, quartz, turquoise, and emerald.

Hot water can also cook up collections of valuable metals along cracks, like gold and copper. And it even helped grow the giant, milky-white crystals

of gypsum found in Mexico's Cave of Crystals. These astounding spikes are taller than a person! They grew underground from water that had been heated by a deep blob of magma.

Hot water also helps plain-looking rocks called geodes hide dazzling secrets. Crack open a geode and you'll find a pocket full of beautiful crystals inside! The minerals grow in holes left behind by bubbles of gas that got stuck in volcanic rocks. When hot water fills the gaps, minerals can grow, like purple amethyst or fluffy white okenite.

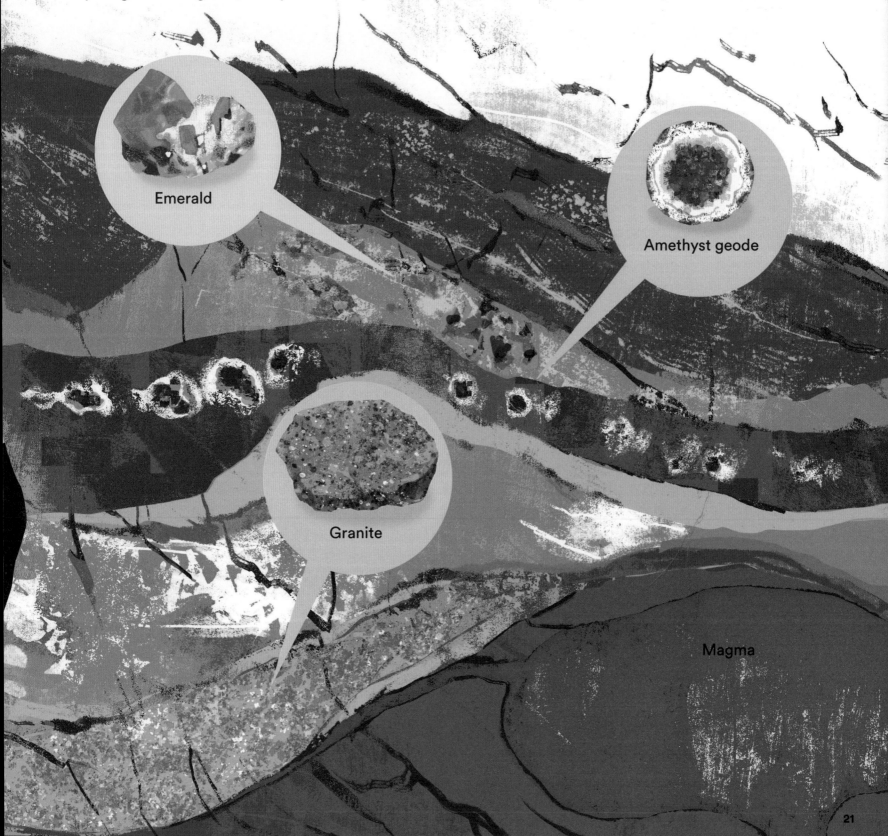

Emerald

Amethyst geode

Granite

Magma

Some rocks tell stories of crashes and collisions

Are you standing still? Yes? Think again! Whether or not you can feel it, the land under your feet is always creeping along—but don't worry, it's only moving about as fast as your fingernails grow.

The land moves because Earth's crust is broken into big pieces of rock, known as tectonic plates. These huge plates are always moving and fit together like a jiggly jigsaw.

Sometimes the plates push together so hard they "slip," or move quickly. This can cause the ground to shake and crack in an earthquake—even underwater.

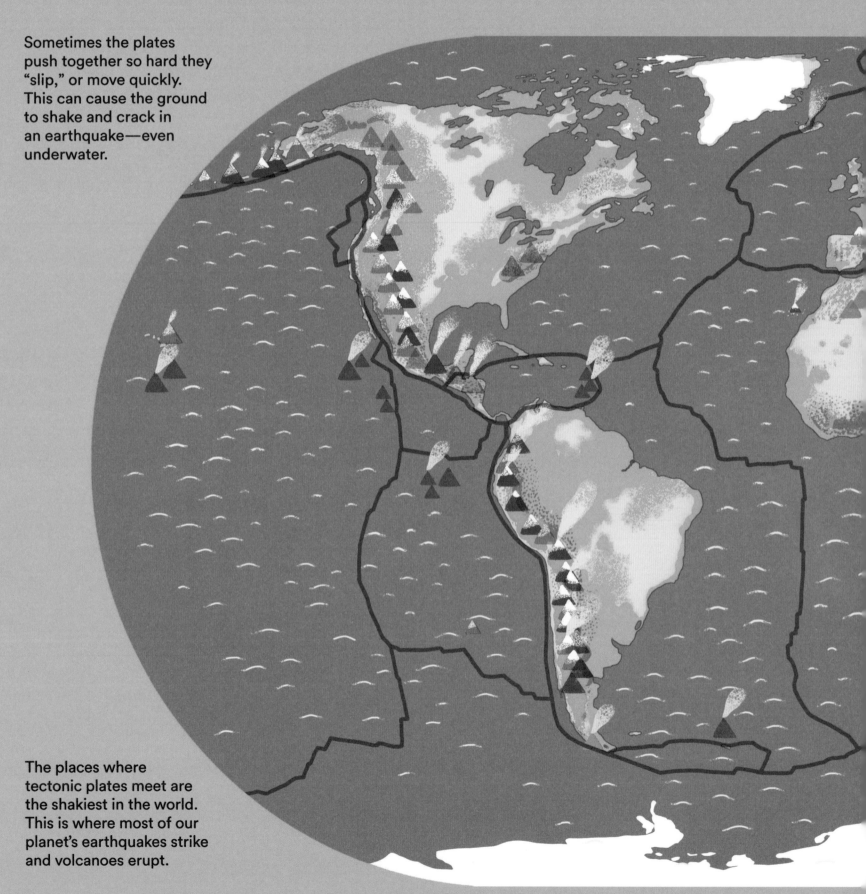

The places where tectonic plates meet are the shakiest in the world. This is where most of our planet's earthquakes strike and volcanoes erupt.

METAMORPHIC ROCKS

If you find a rock with stripes, squiggles, or even flaky layers, it may have formed when two tectonic plates pushed together, smooshing and smearing rocks deep underground. All this mashing can create a new rock type called a metamorphic rock. Doesn't this one look nice? That's also how you say its name—gneiss.

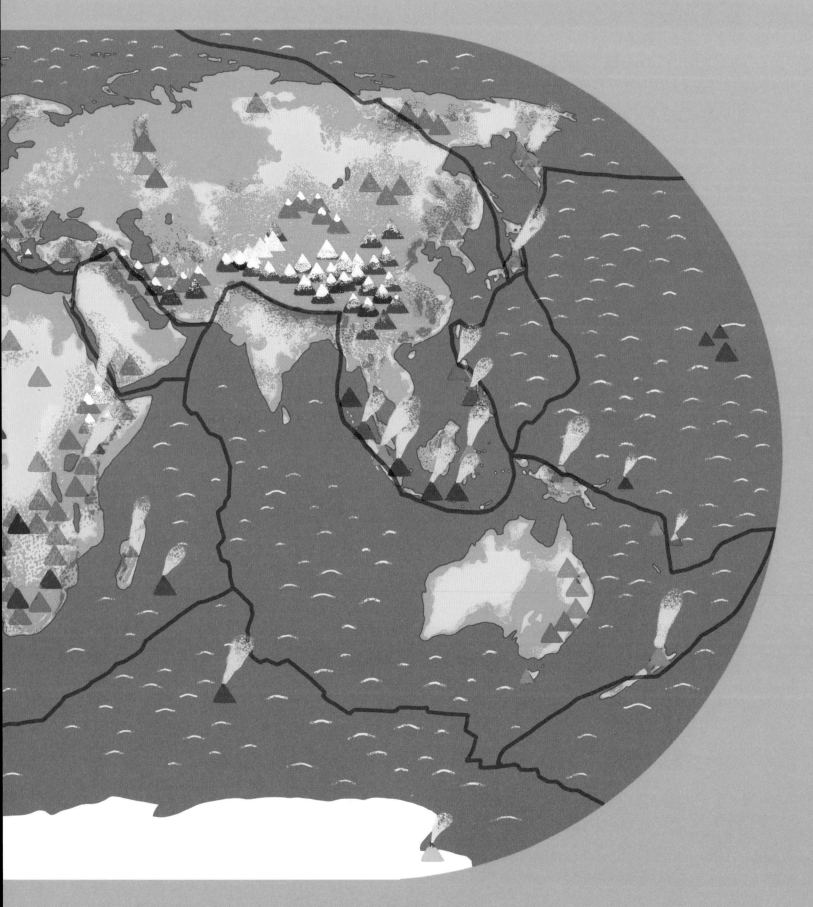

Some rocks tell how mountains grow

Did you know that you can find seashells on top of a mountain? Believe it or not, you can! But how would something from the ocean end up miles away, high above the clouds? The answer can help tell us how mountains grow.

When tectonic plates crash into each other, big things can happen. The crash slowly pushes

Broken bits of ancient ocean critters found on Mount Everest include trilobites, which were one of the first ancient creatures with eyes. You can also find bits of shells from brachiopods. These critters might look like scallops on the outside, but hiding in their shells is a crown of wiggling tentacles that sweep food into their mouths.

great blocks of land up into mountains that rise high into the sky. If the land once had an ocean, the crash would push bits of seafloor up, up, up higher than where birds fly.

That's what happened at Mount Everest, the world's highest mountain. Hikers who make it to the peak are in for a treat: they can find broken bits of ancient shells in rocks that once sat under the sea.

Many of these mountain-making crashes are still happening. So Everest might still be growing. Super slowly, of course.

HOW MOUNTAINS GROW
As two continents move closer, the ocean between them shrinks. After thousands or millions of years, the blocks of land finally crash together, crumpling the rock like a piece of paper, draining the ocean, and shoving mountain peaks sky high.

Some rocks reveal adventures
from mountain to sea

While stones might seem tough, they don't last forever. As mountains grow, wind and water also wear them down, causing some rocks to crumble. Rain sweeps the rocky rubble into rivers, carrying the stony bits down steep slopes.

Rivers are nature's greatest rock collectors—big rocks, little rocks, red rocks, tan rocks. As the water flows downstream, these rocky bits tumble and break into smaller pieces, becoming little pebbles, gravel, and even grains of sand. Smaller streams join the rushing river, feeding its waters and drawing in rocks from near and far.

If you follow a river long enough, you'll often end up at the sea. This is where many small stony bits collect in sandy beaches after their big journey.

Sand tells stony stories too

What color is sand? You might say it's yellow, tan, or white. But sand can be almost any color of the rainbow. Each color is a clue to the rocks it formed from.

Beaches with yellow or tan sand are super common around the world. That's because sand of these colors is mostly made from quartz, which is a mineral found in many rocks. Other colors of sand are made from different minerals, like red garnet or green olivine.

The sand's size and shape also help tell its story. The tiniest and roundest sand grains have probably traveled the farthest. You might also find a few small surprises hiding in sand, like broken bits of shells, coral, or the star-shaped skeletons of tiny creatures washed in by ocean waves.

What colors and shapes of sand have you seen?

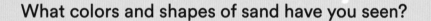

The black-sand beaches in places like Iceland, Hawai'i, and Spain's Canary Islands are the worn-down and broken bits of dark basalt rocks that erupted from volcanoes.

Some beaches in Namibia are streaked with purple-pink sand. These grains were formed from larger rocks that contained a deep-red mineral known as garnet.

Some Hawaiian beaches are full of green sand. These pretty grains come from basalt rocks that are packed with the bright green mineral olivine.

A rock's layers are like the pages of a book

Sand isn't always where a rock's adventure ends. It's often just another stop in its many travels. When sand sinks to the bottom of lakes and oceans it forms layers like a stack of blankets. These layers can be many colors—red or orange, gray or even pink!

SEDIMENTARY ROCKS

If you find a stone that looks like it's made from tiny grains of sand or pieces of rock, you probably have a sedimentary rock. Each layer of the rock can tell you what life was like on Earth at different times, long ago. It's like reading a book with a new story on each page!

Each new layer sits on top of an older one, pressing down on everything below until something magic happens: the tiny bits of rock turn into a new stone. These are called sedimentary rocks.

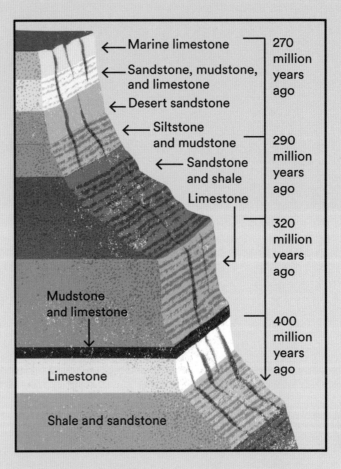

← Marine limestone	270 million years ago
← Sandstone, mudstone, and limestone	
← Desert sandstone	
← Siltstone and mudstone	290 million years ago
← Sandstone and shale	
Limestone	320 million years ago
Mudstone and limestone	400 million years ago
Limestone	
Shale and sandstone	

A CLOCK IN ROCK

Different types of rock can also stack on top of each other, which is part of what gives Arizona's Grand Canyon its thick, colorful bands of red, orange, brown, and tan. Each type of rock, like limestone or sandstone, gives us clues about what life was like on Earth when each layer formed. The Grand Canyon's top layer of limestone was once a seafloor.

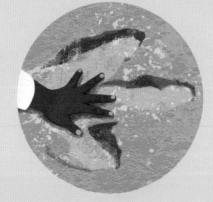

TREASURES IN LAYERS

Sedimentary rocks can reveal all kinds of Earth's old secrets. Its layers can show ripples from passing waves, dimples from ancient raindrops, or even footprints of creatures that once stomped on the land long ago. In the Grand Canyon, these sandy records include the imprint of a giant dragonfly wing!

What can rocks reveal about ancient animals?

Millions of years ago, dinosaurs ruled the land, spiral-shelled ammonites glided through seas, and winged reptiles called pterosaurs soared across skies. But how do we know what they looked like? Rocks, of course!

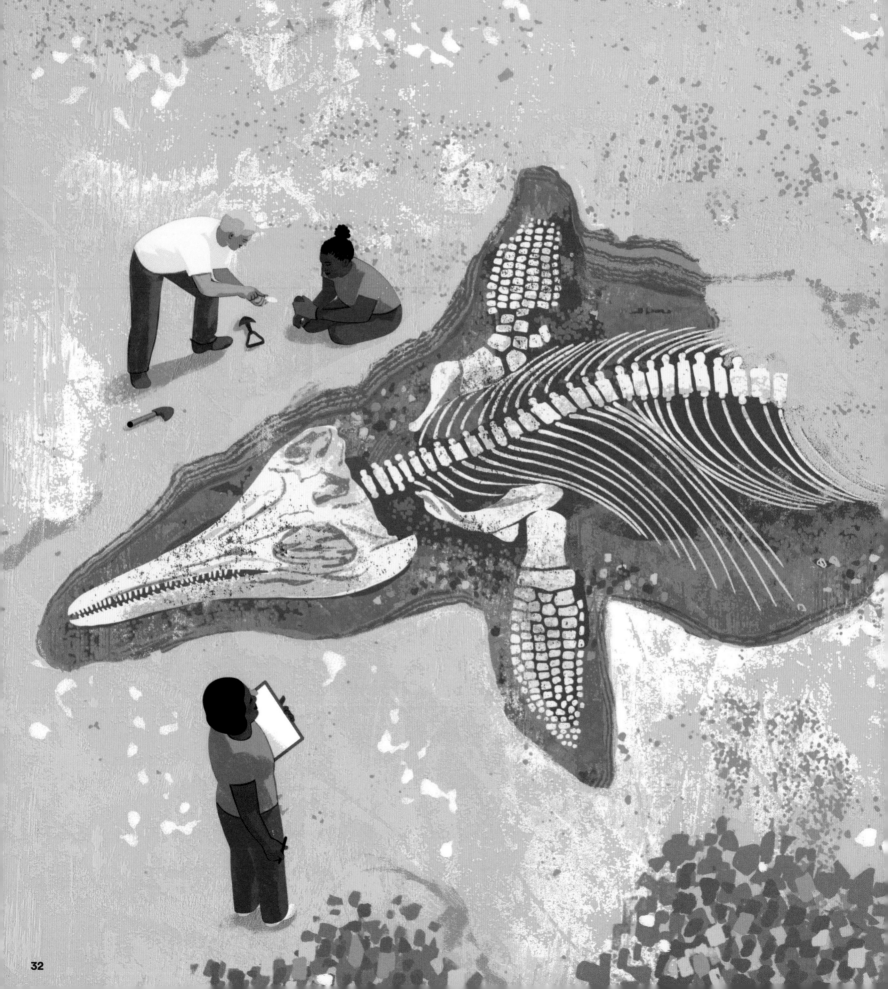

Many clues come from fossils, which are ancient plants or animals turned to stone. They form when a critter dies and sand or mud quickly buries their bodies.

Millions of years pass as layer after layer of sand and rocks stack on top of the ancient critter. And trickling water carries minerals that turn their bones to stone.

Scientists called paleontologists have found fossils on every continent, even Antarctica. These rocks tell us all kinds of stories about ancient life—enormous insects buzzing through skies, feathered dinosaurs lumbering on land, and mushrooms towering as tall as trees.

This fossil is an ancient dolphinlike creature called an ichthyosaur, which swam in the seas while dinosaurs roamed the land. Ichthyosaurs could be gigantic—some were longer than 70 feet! That's about as long as two buses lined up end to end.

What can rocks reveal about Earth's oldest living thing?

Some of the oldest fossils ever found give a slippery clue: early life probably grew in slimy, gooey mounds.

Before there were animals and bugs, or trees, flowers, and grass, bacteria ruled the world—and they're still around today. These tiny living creatures are so small you can't see them with

If you cut one of these slimy mounds in half, you would see the crinkly layers made by the bacteria. Stromatolite fossils that are millions of years old can still have olyers!

your eyes. But when many bacteria live together they can form large layers of slime!

When Earth was just a baby, these gooey layers grew in shallow ocean waters. Waves washed over them and sand stuck to the slime. So new bacteria would grow a fresh slimy layer on top. This happened again and again until the goo grew into a mound that we call a stromatolite.

You can visit living relatives of these ancient slimy mounds on the coast of western Australia, where bacteria are still growing layer by layer just like their relatives did billions of years ago.

Some rocks tell stories about space

On a clear, dark night, look up. You might see a flash of light across the sky. What do you think it is? These are often called shooting stars, but they're not really stars. Each one is actually a glowing rock from space that's falling to Earth!

What makes these space rocks glow? Earth is wrapped in a layer of gases, called the atmosphere, which protects our planet and includes the air you breathe. But when space rocks enter the atmosphere, falling super fast, they get so hot they start to glow. Sometimes these rocks burn up and disappear, but if one survives the fall it's called a meteorite.

Most space rocks are extremely old. They started as a dusty cloud swirling around the Sun when it was just a baby—our planet hadn't even formed yet. Over time, these clouds of dust stuck together, making planets and even growing into our own planet, Earth. Studying meteorites helps reveal the stories of how the Sun and planets formed.

METEORITES

Most meteorites are made of—you guessed it—rock! One type of stony meteorite called chondrites are the oldest rocks we've ever found. There are also iron meteorites, which are made of metal and might be like our own planet's metal core. There are also stony-iron meteorites, which may be the prettiest space rocks you'll ever see. One type called pallasites are lumps of metal speckled with sparkling green crystals of the mineral olivine.

What do moon rocks reveal?

When our planet was born, it looked nothing like it does today. There were no houses or people, no forests or animals, no oceans or fish. But can you guess what there was? That's right—rocks!

At first, Earth was so hot that the rocks were just a goopy soup. And we're not completely sure what happened next. Our planet's restless surface has erased most rocky clues from that long ago.

But there is one big clue that tells us about Earth's early days: the Moon.

When Earth was young, many big space rocks whizzed by. One of these space rocks, a big planet-sized one, probably crashed into baby Earth, blasting out a huge cloud of dust and rocks. The cloud slowly swirled and smooshed together until it formed the rocky ball that rises in the sky every night—it's the Moon!

So the next time you look up at the Moon, just remember you're gazing at a rock made from bits of baby Earth.

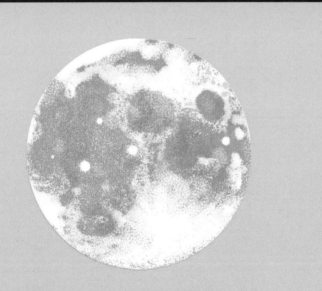

THE MOON'S SPECKLED FACE

Thousands of space rocks keep crashing into the Moon, making big dents on its bright face. From Earth, they look like polka dots. Scientists think the rocks of these moon craters could tell us more about ancient space rocks that might have crashed to Earth and brought water or even the ingredients for life.

What do we know about rocks on other worlds?

The Moon is the farthest humans have ever traveled in space. But we've sent robots much farther. Right now, two robots with wheels—called rovers—are hunting the surface of Mars for hints of ancient life and clues to help us understand the red planet's past.

Today Mars is cold and dry, but the rocks hold hints that Martian rivers once flowed across the ground. On Earth, rivers carve channels, dumping the scooped-out sand and gravel in fan shapes at the edges of lakes and seas. Mars has similar channels and fans, but they're all dried up. Maybe—just maybe—water once flowed on Mars and alien life hid in the muck.

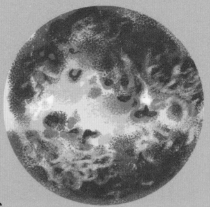

VENUS

Earth is not the only place with an explosive past. The planet closest to us, Venus, has tons of volcanoes. Scientists recently counted about 85,000 volcanoes on Venus —and some may even still flow with lava.

ENCELADUS

Saturn's moon Enceladus looks like a ball of ice. But there are probably lots of rocks beneath this chilly outer layer. Scientists have even spotted water and tiny bits of rock spurt through cracks in the ice.

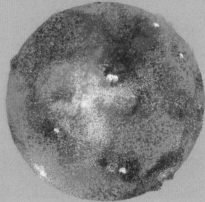

CERES

This oddball is a dwarf planet, which means it's smaller than other planets. Ceres probably has a rocky core but is also covered in ice. And ice mountains on Ceres may erupt like volcanoes but spit out frosty water rather than lava!

Scientific detectives

What are geologists? They're scientists who study the rocks of our planet and beyond in space. Like detectives, they sort through clues left in rocks about things that happened long ago, or sometimes just yesterday. Rocks can also help geologists figure out what Earth might look like millions of years from now.

To solve these planetary puzzles, geologists travel around the world, climbing mountains, sailing across oceans, and hiking through deserts and jungles.

DIGGING DEEP

Some geologists search for clues to our planet's past in long skinny columns of sand or rock, called cores. They collect these columns from soft layers at the bottom of lakes or drill them out of the rocky seafloor.

SHAKING IT UP

Seismologists use special machines called seismometers to record every time our planet shakes from an earthquake or volcanic rumble. This helps us understand where the ground might tremble in the future or when volcanoes might erupt.

DRILLING DOWN

Glaciologists study some of the coldest places on Earth, where ice collects in big chunks called glaciers. By drilling cores of glacier ice, they can study what Earth's climate was like in ancient times and figure out how it will change in the future.

THIRST FOR KNOWLEDGE

Hydrologists study how water flows and where it collects on our planet, both above and below ground. Some hydrologists test the chemistry of water to ensure it's safe to drink.

EXPLOSIVE CURIOSITY

Volcanologists collect samples of volcanic ash, rock, or even scorching hot lava to help us understand how and why these fiery peaks erupt.

There are many questions scientists have yet to answer. Perhaps you could help!

Rocks in our lives

Geologists also study rocks to find minerals and other hidden treasures that we can use in our daily lives, like copper wires that help light our homes or iron to make the cars we drive. Rocks may also be beautiful and valuable, like sparkly gems in jewelry. When gemstones are found in nature, they're often hidden in other types of rock or have rough or broken edges. A jeweler cuts and polishes them so they glimmer and gleam.

Jewelry Box

Limestone tiles

Slate wall

Pebble path

Slate floor

Marble countertop

Rocks themselves can also be very useful. Look around your home—how many uses of rock do you see?

Perhaps you can see the colorful speckles of granite or white-and-gray swirls of marble on kitchen counters or bathroom walls. Even glass is made from minerals!

Clay bricks

Mineral collection

Glass window

Metal lamp

Shale stepping stones

Earth is full of wild and wonderful rocks

Rocks are everywhere—but in many places they're extra gorgeous and grand! If you visit these amazing sights and spot a special stone, don't forget to check if you can collect before taking any treasures home. Many parks want the rocks to stay in place, so that everyone has a chance to learn their stony secrets.

SALAR DE UYUNI, BOLIVIA
About 40,000 years ago, this area was covered by giant lakes. But the water slowly evaporated, leaving behind a thick layer of salt.

METEOR CRATER, ARIZONA, UNITED STATES
About 50,000 years ago a giant iron meteorite slammed into Earth's surface here, blasting out a crater three-quarters of a mile long.

POSTOJNA CAVE, SLOVENIA
An underground river carved this cave millions of years ago as it rushed through rock. Then tiny trickles of mineral-rich water helped create the spikes of stone that poke down from the ceiling and up from the floor.

ULURU, NORTHERN TERRITORY, AUSTRALIA
This tower of sandstone formed around 500 million years ago as tectonic crashes shaped Australia. The red color comes from iron in the rocks that rusted like a bike left in the rain.

GIANT'S CAUSEWAY, IRELAND

These pillars of rock formed as an ancient flood of lava cooled and cracked into the five- and six-sided shapes that make the big steps of the causeway.

U-SHAPED VALLEYS OF FIORDLAND, NEW ZEALAND

As glaciers swept through this region long ago, they ground down the rocks below, carving valleys into smooth swoops shaped like giant letter U's.

Australia's Indigenous people consider Uluru sacred, created by ancestral beings at the beginning of time. They believe the spirits of these ancestors still linger in the rocks.

THE RAINBOW MOUNTAINS, ZHANGYE NATIONAL GEOPARK, CHINA

Water helped lay down colorful sand and minerals that were later pressed into rock. Tectonic shifts tilted the land to form the stunning mountains you see today.

Author's note

When I was in elementary school, I loved to sit on the playground and search through the gravel for the prettiest pieces. I found shiny yellow rocks, bright pink rocks, and even clear rocks. These little bits of gravel were the very first members of my rock collection.

Growing up, I loved anything that sparkled. (Geodes? Yes please!). But one of my favorite things to find were tiny round bits of fossilized crinoid, an ancient marine creature that looked a little like a feather duster while alive. All these rocks felt magical to me, each one a piece of our planet's past. And the older I grew, the more I discovered what rocks could reveal about the world around us.

Geology is a huge subject and covers everything from Earth's hot metallic core to the tip of its tallest peaks and beyond to other worlds. There are also many things about Earth that scientists still don't fully understand—and many rocky clues to these mysteries that they have yet to find. Maybe you'll be the one to discover what secrets still lie deep beneath our feet!

Though I could only include a small part of the gigantic world of geology, I hope this book inspires you to take a closer look at every boulder, pebble, piece of gravel, and grain of sand you see and think about the amazing stories hiding in every stone.